Patterns Across the Cu

MW01101150

Table of Cont

Introduction 1
General Assessment 2

Unit 1: Mathematics
Unit 1 Assessment 4
Finding a Pattern 6
Color Patterns 7
Shape Patterns 8
Size Patterns 9
Ordering Numbers 10
Number Patterns 11
Ordinal Numbers 12
Counting On 13
Doubles Facts 14
Order in Addition 15
Adding 0 16
Subtracting 0 17
Fact Families 18
Even and Odd Numbers 19
Same Sum Patterns 20
Numbers to 100 21
Skip Counting by 10s 22
Skip Counting by 2s 23
Skip Counting by 5s 24
Time to the Hour 25
Using the Calendar 26
Designing a Pattern 27

Unit 2: Language Arts
Unit 2 Assessment 28
ABC Order 29
Vowel-Consonant-**e** Spelling Pattern 30
Naming One and More Than One 31
Verbs in the Present 32
Poetry Patterns 33

Unit 3: Social Studies
Unit 3 Assessment 34
Leaders 35
Map Keys 36
Directions 37
Making a Product 38
Counting Money 39

Unit 4: Science
Unit 4 Assessment 40
People Grow and Change 41
Plants Grow and Change 42
Grouping Animals 43
Dressing for the Season 44
Weather Patterns 45

Student Progress Chart 46
Answer Key 47

Introduction

The world is full of patterns. House numbers form a pattern, words are spelled according to patterns, and leaves on a tree branch form patterns. The identification of patterns is a beginning foundation that leads to the understanding of algebra. According to the National Council of Teachers of Mathematics, students "should analyze the structure of the pattern and how it grows or changes, organize this information systematically, and use their analysis to develop generalizations about the mathematical relationships in the pattern." (*Principles and Standards for School Mathematics*, page 159.) *Patterns Across the Curriculum* engages students in identifying relationships applicable in all content areas. The book is divided into four units: Mathematics, Language Arts, Social Studies, and Science. Intriguing content-area connections will help students enrich and extend their understanding of relationships in all aspects of their lives.

Unit 1: Mathematics
The pages in this unit identify patterns found in mathematics, such as number, color, and shape patterns, addition strategies, and time to the hour.

Unit 2: Language Arts
Here, students examine ABC order, long vowel words, and present tense verbs.

Unit 3: Social Studies
Students explore directions, map keys, and the role of leaders in groups.

Unit 4: Science
In this unit, students look at such patterns as the growth cycles of living things, weather patterns, and clothing patterns as related to seasons.

Notes
Assessment
There are two kinds of assessments.
- On pages 2 and 3 is a general assessment that covers important patterns appropriate for the first grade. It can be given as a pretest to gauge students' knowledge of patterns. Later in the year, the same test can be administered to determine students' understanding, progress, and achievement.
- Each unit also has an assessment. These can be administered at any time during the unit as a pretest, review, or posttest for specific concepts.

Student Progress Chart
Duplicate the grid sheet found on page 46. Record student names in the left column. Note the date of completion of each lesson for each student.

Name _____   Date _____

General Assessment

★✧★✧ **Complete the pattern.**

1. ● ■ ● ■ ● ■ _____

2. ▲ ■ ■ ▲ ■ ■ ▲ _____

3. • • ● ● ● ● _____

4. 8 7 6 5 4 3 _____

5. 2   4   _____   _____   10

★✧★✧ **Add or subtract.**

6. 5 + 2 = _____   7. 0 + 8 = _____

8. 4 – 0 = _____   9. 3 – 3 = _____

Go on to the next page.

General Assessment, p. 2

⭐✧⭐✧ **Write the letters in ABC order.**

10. h   i   g   ___   ___   ___

___   ___   ___

11. v   u   t   ___   ___   ___

___   ___   ___

⭐✧⭐✧ **Write the word to show that the action happens now.**

12. The boy _____.

eat

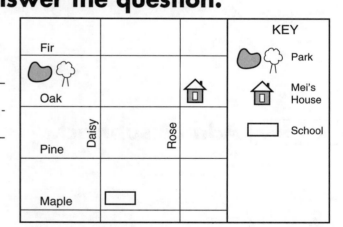

13. The kitten _____.

play

⭐✧⭐✧ **Use the map to answer the question.**

14. What does ▭ mean?

_____

_____

⭐✧⭐✧ **Order the pictures to show how a plant grows. Write 1, 2, 3, and 4.**

15.

___   ___   ___   ___

Unit 1 Assessment

★✧★✧ Complete the pattern.

1. ■ ▲ ■ ▲ ■ ▲ _____ _____

2. ● ● ★ ● ● ★ _____ _____

3. ■ ■ ■ ■ ■ _____

4. 1   2   3   4   5   _____ _____

★✧★✧ Add or subtract.

5. 6 + 2 = _____

6. 6 + 0 = _____

7. 3 − 3 = _____

8. 2 − 0 = _____

Go on to the next page.

Name _____   Date _____

Unit 1 Assessment, p.2

★✧★✧ **Add or subtract to complete the fact family.**

9.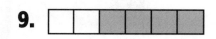

            _____   _____

$2 + 4 =$ _____ $4 + 2 =$ _____

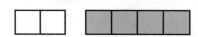

          _____   _____

$6 - 4 =$ _____ $6 - 2 =$ _____

★✧★✧ **Write <u>even</u> or <u>odd</u>.**

10.

_____

- - - - - - - - - - - - - - - - - - - -

11. ● ●
   ● ● ●

_____

- - - - - - - - - - - - - - - - - - - -

★✧★✧ **Complete the skip-count pattern.**

12.   2   _____   6   _____   _____

Finding a Pattern

Look at the circles.
One circle is white.
One circle is dark.
The circles keep going in the same order.
They make a **pattern**.

★✧★✧ **Circle the one that shows a pattern.**

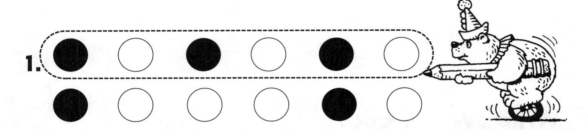

1.

2.

3.

4.

Color Patterns

You can use **colors** to make a pattern.

★✧★✧ **Color the squares.**
Finish the pattern.

1.

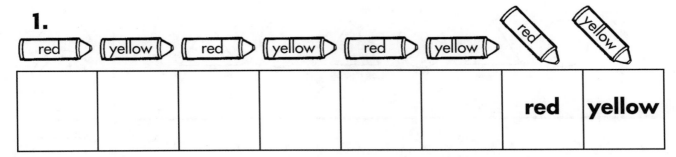

red | yellow | red | yellow | red | yellow | **red** | **yellow**

2.

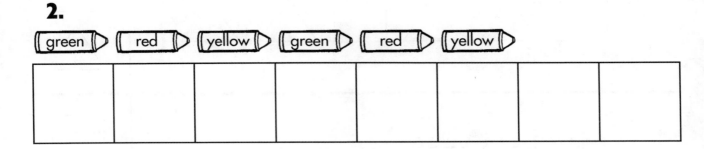

green | red | yellow | green | red | yellow

3.

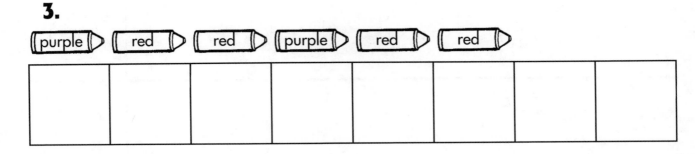

purple | red | red | purple | red | red

4.

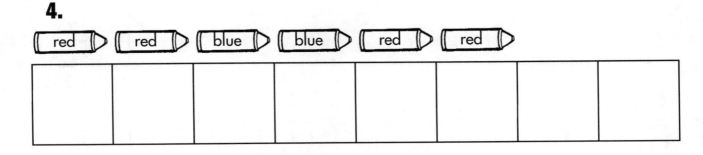

red | red | blue | blue | red | red

Shape Patterns

You can use **shapes** to make a pattern.

★✦★✦ **Circle the one that comes next in the pattern.**

1.
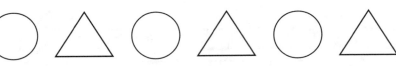

2.

3.

4.
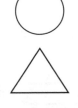

5.
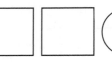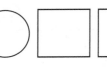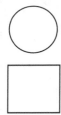

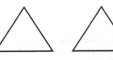

Size Patterns

You can use the **size** of things to make a pattern.

★✧★✧ **Draw the one that comes next in the pattern.**

1.

2.

3.

4.

★✧★✧ **Cut out the squares.**
Order them from small to large.
Paste them on paper to make a pattern.

Ordering Numbers

These numbers are in order.
1 2 3 4 5 6 7 8 9 10

★✧★✧ **Write the numbers in order.**
Which is missing?

1. 7 8 6 4   4 6 7 8   5

2. 9 5 8 6

3. 6 2 4 3

4. 3 7 4 6

5. 1 5 4 2

6. 7 9 6 10

7. 8 5 9 6

8. 5 4 7 8

Number Patterns

Numbers are written in order to make a pattern.
Numbers can be written out of order to make a pattern, too.

★✧★✧ **Write the numbers that come next in the pattern.**

1.

2.

3.

4.

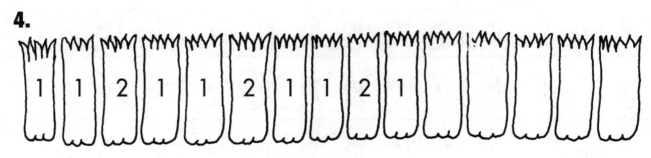

5.

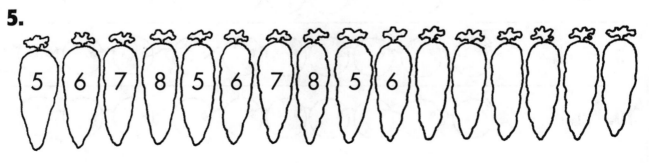

Name _____   Date _____

Ordinal Numbers

Ordinal numbers name a place.

first   second   third   fourth   fifth   sixth   seventh   eighth   ninth   tenth

★✧★✧ **Color to show the order.**

1. first

2. second

3. third

4. fourth

5. fifth

6. sixth

7. seventh

8. eighth

9. ninth

10. tenth

Unit 1: Mathematics
Patterns Across the Curriculum 1, SV 3424-X

Counting On

Look at the addends. Is a number 1, 2, or 3?
Count on to add.

$2 + 2 = \square$

$2 + 2 = 4$

★✧★✧ **Count on to add. Write the sum.**

1.

$5 + 1 = \underline{\quad}$

2.

$3 + 2 = \underline{\quad}$

3.

$7 + 2 = \underline{\quad}$

4.

$4 + 3 = \underline{\quad}$

Unit 1: Mathematics
Patterns Across the Curriculum 1, SV 3424-X

Doubles Facts

A **doubles fact** adds the same addends.

 **Make a double.**
Draw the same number of dots.
Complete the doubles fact.

1.

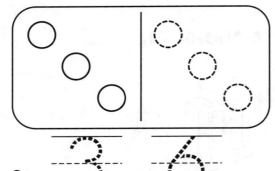

3 + _3_ = _6_

2.

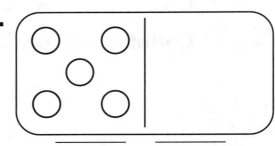

5 + _____ = _____

3.

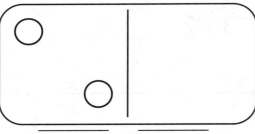

2 + _____ = _____

4.

4 + _____ = _____

5.

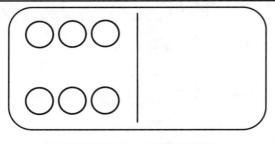

6 + _____ = _____

6.

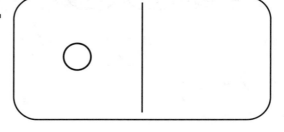

1 + _____ = _____

Order in Addition

Numbers may be added in any order.
The sum will stay the same.

 Draw X to show how many.
Write the sum.

1.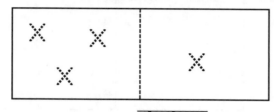

3 + 1 = **4**   1 + 3 = _____

2.    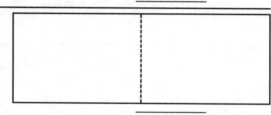

2 + 4 = _____   4 + 2 = _____

3. 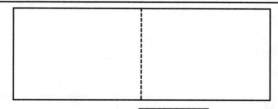   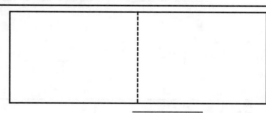

4 + 1 = _____   1 + 4 = _____

4.    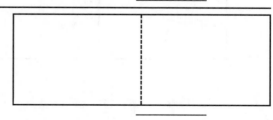

3 + 2 = _____   2 + 3 = _____

Adding 0

| Any number plus **0** equals the same number. |

★✧★✧ **Draw circles to show how many. Write the sum.**

1.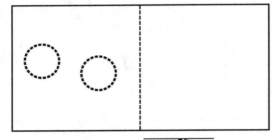

$2 + 0 = 2$

2.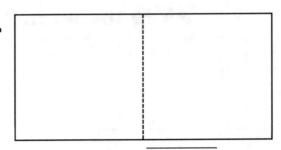

$0 + 3 = $ _____

3.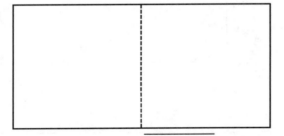

$0 + 5 = $ _____

4.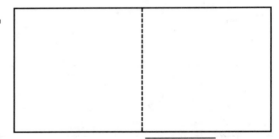

$1 + 0 = $ _____

5.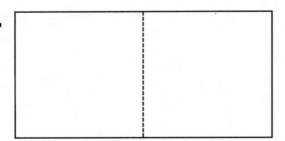

$6 + 0 = $ _____

6.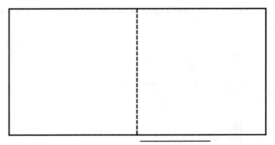

$0 + 4 = $ _____

Name _____   Date _____

Subtracting 0

If you subtract same numbers, all are taken away.

$4 - 4 = 0$

If you subtract 0, none are taken away.

$4 - 0 = 4$

★✧★✧ **Cross out birds to show how many are taken away. Subtract.**

1.

$3 - 3 =$ _____

2.

$3 - 0 =$ _____

3.

$5 - 0 =$ _____

4.

$5 - 5 =$ _____

5.

$7 - 7 =$ _____

6.

$7 - 0 =$ _____

Fact Families

Look at the number sentences.
They belong in the same **fact family**.
A fact family uses 3 numbers.

3 + 2 = 5   5 – 2 = 3

2 + 3 = 5   5 – 3 = 2

2, 3, and 5 are the numbers in this **fact family**.

★✧★✧ **Add or subtract.**

1.

5 + 2 = _____   7 – 2 = _____

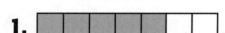   

2 + 5 = _____   7 – 5 = _____

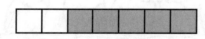   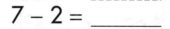

2.

7 + 1 = _____   8 – 1 = _____

1 + 7 = _____   8 – 7 = _____

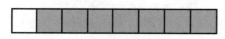   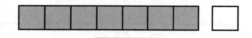

Name _____ Date _____

Even and Odd Numbers

Dots can form patterns to show a number.

4

3

Even numbers have none left over. The number 4 is **even**.

Odd numbers have 1 left over. The number 3 is **odd**.

★✧★✧ **Circle pairs of dots. Write <u>even</u> or <u>odd</u>.**

1.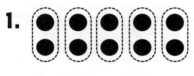

_____

- - - - - - - - - - - - - - - - - - -

_____

2.

_____

- - - - - - - - - - - - - - - - - - -

_____

3.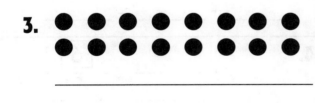

_____

- - - - - - - - - - - - - - - - - - -

_____

4.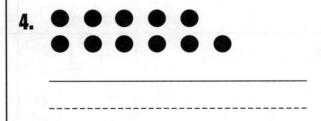

_____

- - - - - - - - - - - - - - - - - - -

_____

5.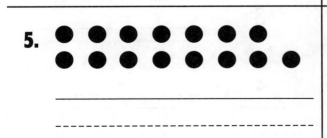

_____

- - - - - - - - - - - - - - - - - - -

_____

6.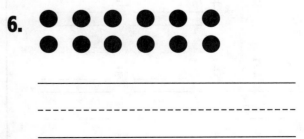

_____

- - - - - - - - - - - - - - - - - - -

_____

Unit 1: Mathematics
Patterns Across the Curriculum 1, SV 3424-X

Same Sum Patterns

Look at the addends.
Do you see a pattern?

3 + 2 = 5
2 + 3 = 5
1 + 4 = 5
0 + 5 = 5

★✧★✧ **Complete the addition sentences.**

1. 9

5 + __4__ = __9__

6 + _____ = _____

7 + _____ = _____

_____ + _____ = _____

2. 8

4 + _____ = _____

5 + _____ = _____

6 + _____ = _____

_____ + _____ = _____

3. 7

3 + _____ = _____

4 + _____ = _____

5 + _____ = _____

_____ + _____ = _____

4. 6

2 + _____ = _____

3 + _____ = _____

4 + _____ = _____

_____ + _____ = _____

Numbers to 100

Numbers on a 100s table make different patterns.

★✧★✧ **Complete the table.**
Write the missing numbers.

1	2	3	4	5	6	7	8	9	10
11		13	14	15		17	18		20
21		23		25	26		28		30
31		33	34		36	37		39	
41									
51		53	54	55		57		59	60
61		63	64		66			69	70
71		73		75			78	79	
81		83	84		86	87			90
91		93		95	96		98		100

★✧★✧ **Tell about a number pattern you see.**

_____

- -

_____

_____

- -

_____

Skip Counting by 10s

Sometimes, you count and leave out numbers. You are **skip counting**.

★✧★✧ **Skip count by 10s. Write how many.**

1.

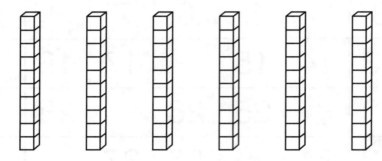

2.

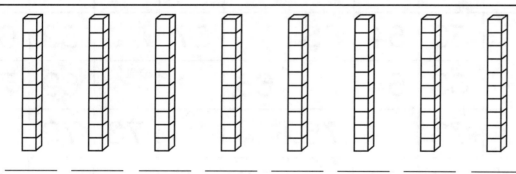

3.

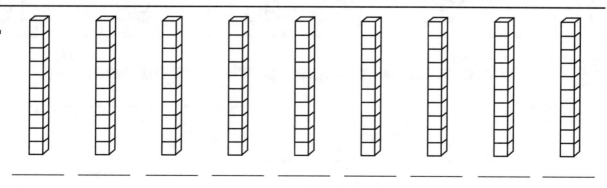

Skip Counting by 2s

Sometimes, you count and leave out numbers.
You are **skip counting**.

★✧★✧ **Skip count by 2s.**
Complete the pattern.

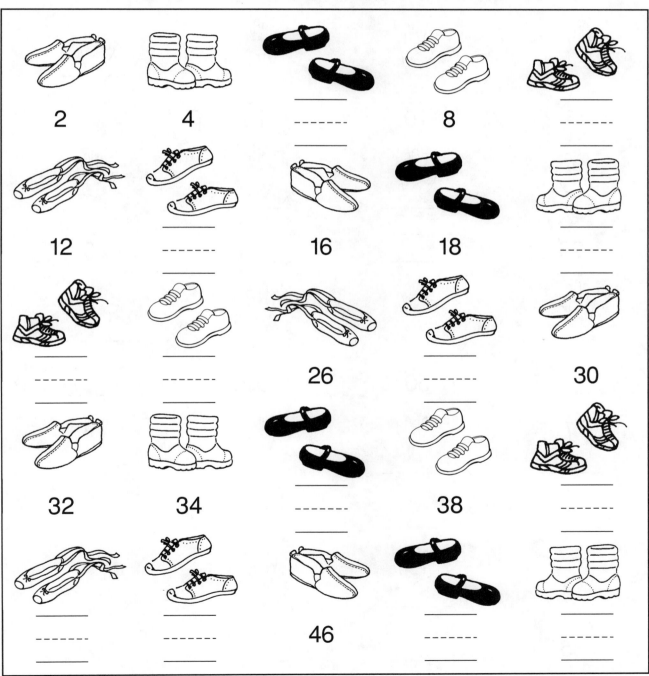

2	4	-------	8	-------
12	-------	16	18	-------
-------	-------	26	-------	30
32	34	-------	38	-------
-------	-------	46	-------	-------

Skip Counting by 5s

Sometimes, you count and leave out numbers. You are **skip counting**.

★✧★✧ **Skip count by 5s. Complete the pattern.**

5   10   15   _____

25   _____   _____   40

_____   50   55   _____

_____   _____   75   _____

85   _____   _____   100

Name _____   Date _____

Time to the Hour

The hands on a clock move in a pattern.
You can tell time to the hour.
The minute hand points to 12.
The hour hand points to another number.

★✧★✧ **Draw the hour hand so that both clocks show the same time.**

1.

`1:00`

2.

`5:00`

3.

`11:00`

★✧★✧ **Write the time on the clock so that both clocks show the same time.**

4.
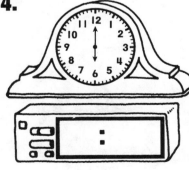

5.

6.
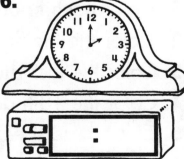

Using the Calendar

> A **calendar** tells many things.
> It tells the month. It tells the days of the week.
> And it tells the dates.

★✧★✧ **Complete the calendar.**

February

Sunday	Monday	Tuesday	Wednesday	Thursday	Friday	Saturday
						1
2		4	5			8
9	10		12	13		15
16		18		20		22
	24		26		28	

★✧★✧ **Find and color these days on the calendar.**

1. Color each Friday red.

2. Color the the first Monday blue.

3. Color the last Thursday yellow.

4. Color the 18th day green.

5. Color these special days orange:
Lincoln's Birthday, February 12,
Washington's Birthday, February 22.

Designing a Pattern

★✧★✧ **Color in some of the shapes to make a pattern.**

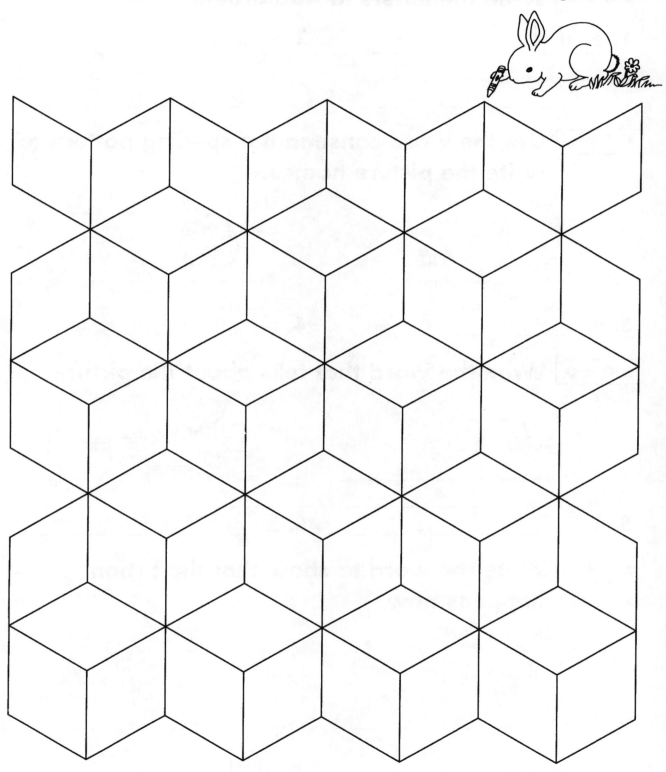

Unit 2 Assessment

★✧★✧ **Write the letters in ABC order.**

1. e   d   f   _____ _____ _____

2. s   q   r   _____ _____ _____

★✧★✧ **Use the word-consonant-e spelling pattern to write the picture names.**

3. v _____ s _____

4. r _____ p _____

★✧★✧ **Write the word that tells about the picture.**

5. 1 _____

6. 2 _____

★✧★✧ **Write the word to show that the action happens now.**

7. The girl _____ .

jump

8. The boy _____ .

run

ABC Order

The order of letters from A to Z is called ABC order.

a b c d e f g h i j k l m n o p q r s t u v w x y z

★✧★✧ **Write the letters in ABC order.**

1. f h g

_____ _____ _____

- - - - - -   - - - - - -   - - - - - -

_____ _____ _____

2. p n o

_____ _____ _____

- - - - - -   - - - - - -   - - - - - -

_____ _____ _____

3. c e d

_____ _____ _____

- - - - - -   - - - - - -   - - - - - -

_____ _____ _____

4. l k j

_____ _____ _____

- - - - - -   - - - - - -   - - - - - -

_____ _____ _____

5. s q r

_____ _____ _____

- - - - - -   - - - - - -   - - - - - -

_____ _____ _____

6. y x w

_____ _____ _____

- - - - - -   - - - - - -   - - - - - -

_____ _____ _____

Vowel-Consonant-e Spelling Pattern

Look at the words **game** and **bone**.
They have a long vowel sound.
They follow a spelling pattern.
The pattern is **vowel-consonant-e**.

cake

rose

★✧★✧ **Use the vowel-consonant-e spelling pattern. Write the picture names.**

1.

b ____ k ____

2.

t ____ p ____

3.

t ____ b ____

4.

r ____ b ____

5.

r ____ k ____

6.

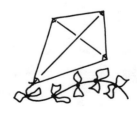

k ____ t ____

Naming One and More Than One

Some naming words tell about one.
Some naming words tell about more than one.
Add an **s** to words that tell about more than one.

one frog

two frogs

★✧★✧ **Write the words that tell about the picture.**

1.

_____

- - - - - - - - - - - - - - - -

_____

2.

_____

- - - - - - - - - - - - - - - -

_____

3.

_____

- - - - - - - - - - - - - - - -

_____

4.

_____

- - - - - - - - - - - - - - - -

_____

Verbs in the Present

Verbs tell about actions.
Verbs can tell about actions that happen now.
Add **s** to an action verb that tells about one person or thing.

The dog **barks**.

★✧★✧ **Write the word to show that the action happens now.**

1.

The fish _____.

swim

2.

The monkey _____.

swing

3.

The boy _____.

play

4.

The girl _____.

talk

Name _____ Date _____

Poetry Patterns

> Words that end with the same sounds are rhyming words. <u>Car</u> and <u>star</u> are rhyming words. Many poems follow a pattern in which the end of the lines rhyme.

Example

Special Things
I love
White snow and blue **bows**.
I love
A sweet red **rose**.
I love
Crunchy sand between my **toes**.
I love
My puppy's wet, black **nose**.

★✧★✧ **Write words from the box to complete the poem.**

mouse   fly   rat

Little Burt Bug is a funny guy.
_____

Sometimes he buzzes like a _____.

Sometimes he meows like a hungry cat.
_____

Sometimes he squeaks like a lonely _____.

But yesterday, asleep in his house,
_____

Burt Bug was as quiet as a _____.

www.svschoolsupply.com
© Steck-Vaughn Company

Unit 2: Language Arts
Patterns Across the Curriculum 1, SV 3424-X

Unit 3 Assessment

★✧★✧ Circle how much money is needed.

1.

★✧★✧ Use the map to answer the questions.

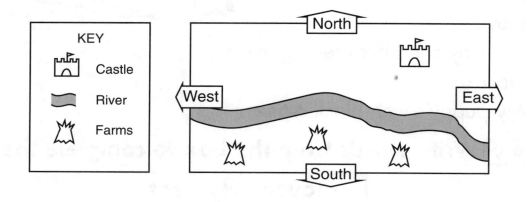

2. What does 🔥 mean? _____

_____

3. In which directions does the river move? _____

4. Draw a farm west of the castle.

Leaders

Most groups have a **leader**.
A leader helps make rules.
A leader makes sure a group follows the rules.
The President is the leader of the United States.

★✧★✧ **Answer the questions.**

1. Who is the principal of your school?

_____

- -

2. How is the principal a leader?

_____

- -

_____

- -

3. Who is the leader in your class?

_____

- -

_____

- -

4. How is the teacher a leader?

_____

- -

_____

Map Keys

A map is a drawing of a place.
Symbols on the map stand for things.
Symbols can be shapes or patterns.
The **map key** tells what the symbols mean.

★✧★✧ **Color the map.**
Draw and color symbols to complete the map key.

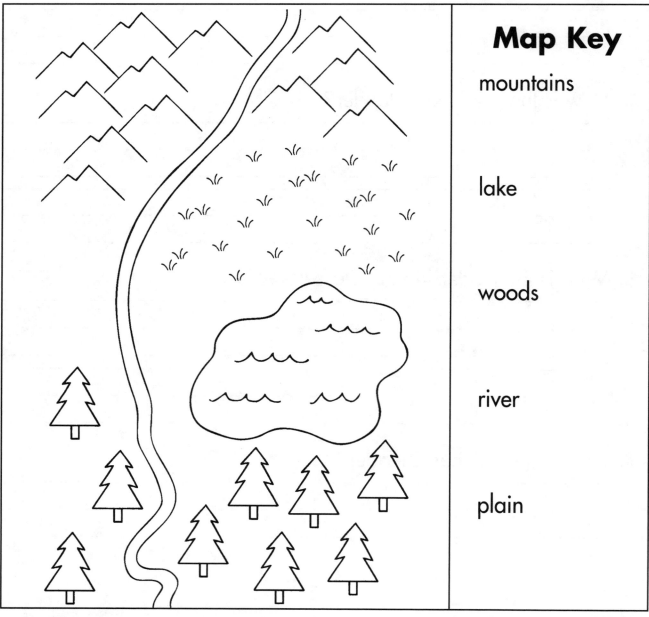

Map Key

mountains

lake

woods

river

plain

Directions

Direction is the path along which something moves.
You can move **north** or **south**.
You can move **east** or **west**.

★✧★✧ Take a ride on the ▶ bus and the ▷ bus.
Write the direction. Write the bus.

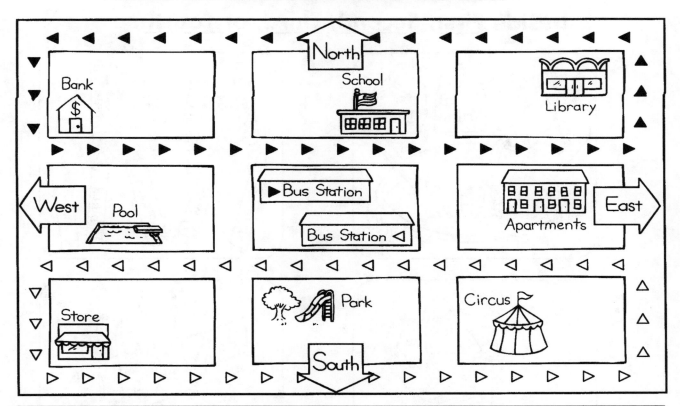

1. 🏦 [East →] to 🏫 on ____▶____ bus

2. 🏪 [→] to 🎪 on _____ bus

3. 📚 [→] to 🏦 on _____ bus

4. 🏊 [→] to 🏪 on _____ bus

Making a Product

A **product** is anything that is made.
Crayons are a product.
Crayons are made the same way each time.

★✧★✧ **How are crayons made?**
Put the pictures in order. Write the letter
beside First, Second, Third, or Fourth.

A

B
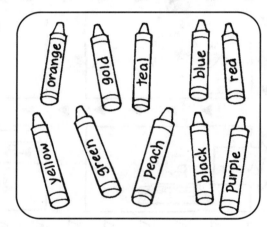

C
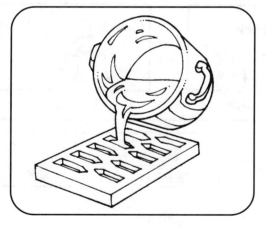

D

First _____   Third _____

Second _____   Fourth _____

Name ______________________ Date ______________

Counting Money

A product is anything that is made.
A toy and a book are products.
People use money to buy products.

penny
1 cent

nickel
5 cents

★✧★✧ Circle how much money is needed.

1. 

2. 

3. 

4. 

Unit 4 Assessment

★✧★✧ **Order the pictures from youngest to oldest.**
Write 1, 2, 3, and 4.

_____   _____   _____   _____

★✧★✧ **Complete the sentences.**
Use words from the box.

clouds   seed   gloves   bird

1. A plant grows from a _____ .

2. A _____ moves by flying.

3. Some people wear _____ in the winter.

4. Big, dark _____ mean it might rain.

People Grow and Change

Living things grow and change. People are living things. People grow and change in a pattern.

★✧★✧ **Cut out the pictures.**
Paste them in order.
Go from the youngest to the oldest.

1.

4.

2.

5.

3.

6.

Plants Grow and Change

Living things grow and change.
Plants are living things.
Plants grow and change in a pattern.

 **Complete the sentences.
Use words from the box.**

seed	flower	root

1. A plant grows from a _____ .

2. A _____ grows under the ground.

3. A _____ makes new seeds.

 **Cut out the pictures. Paste them in order on
paper. Show how a plant grows from a seed.**

Grouping Animals

Animals can be grouped in different ways.
You can group animals by the way they move.

★✧★✧ **Put an <u>F</u> under the animals that fly.
Put a <u>W</u> under the animals that walk.
Put an <u>S</u> under the animals that swim.**

1.

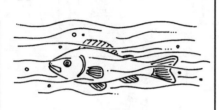

_____

2.

_____

3.

_____

4.

_____

5.

_____

6.

_____

7.

_____

8.

_____

9.

_____

Dressing for the Season

Winter and spring are seasons. Summer and fall are seasons. Seasons change in a pattern. The clothes you wear also change like the seasons.

★✧★✧ Draw lines to match the clothes with the season.

1.

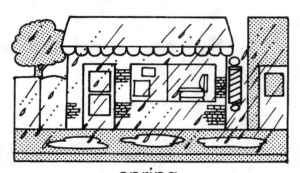

spring

2.

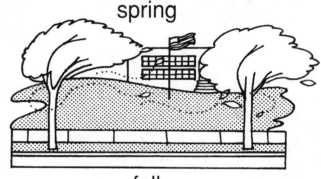

fall

3.

winter

4.

summer

Weather Patterns

> The weather outside makes a pattern.
> Big, dark clouds mean it will probably rain.
> No clouds mean it will be sunny.

★✧★✧ **Cut out the pictures at the bottom of the page.**
Paste each one in the correct box.
Color the pictures.

Student Progress Chart

STUDENT NAME	UNIT 1 MATHEMATICS																									UNIT 2 LANGUAGE ARTS						UNIT 3 SOCIAL STUDIES							UNIT 4 SCIENCE						
	4	5	6	7	8	9	10	11	12	13	14	15	16	17	18	19	20	21	22	23	24	25	26	27	28	29	30	31	32	33	34	35	36	37	38	39	40	41	42	43	44	45

www.svschoolsupply.com
© Steck-Vaughn Company

Patterns Across the Curriculum, Grade 1: Answer Key

p. 2
1. Students draw a circle.
2. Students draw a rectangle.
3. Students draw a larger circle.
4. 2
5. 6, 8
6. 7
7. 8
8. 4
9. 0

p. 3
10. g, h, i
11. t, u, v
12. eats
13. plays
14. school
15. Order: 3, 1, 4, 2

p. 4
1. Students draw a square and a triangle.
2. Students draw 2 circles.
3. Students draw a smaller square.
4. 6, 7
5. 8
6. 6
7. 0
8. 2

p. 5
9. 2 + 4 = 6; 4 + 2 = 6; 6 − 4 = 2; 6 − 2 = 4
10. even
11. odd
12. 4; 8; 10

p. 6
1. Students circle the top pattern.
2. Students circle the top pattern.
3. Students circle the bottom pattern.
4. Students circle the top pattern.

p. 7
1. red, yellow
2. green, red
3. purple, red
4. blue, blue

p. 8
1. circle
2. square
3. circle
4. triangle
5. square

p. 9
1. large square
2. large square
3. large square
4. very small square
For square cutouts, check that students order the squares from small to large.

p. 10
1. 4, 6, 7, 8; missing 5
2. 5, 6, 8, 9; missing 7
3. 2, 3, 4, 6; missing 5
4. 3, 4, 6, 7; missing 5
5. 1, 2, 4, 5; missing 3
6. 6, 7, 9, 10; missing 8
7. 5, 6, 8, 9; missing 7
8. 4, 5, 7, 8; missing 6

p. 11
1. 2, 1, 2, 1, 2
2. 2, 3, 1, 2, 3,
3. 4, 6, 2, 4, 6
4. 1, 2, 1, 1, 2
5. 7, 8, 5, 6, 7, 8

p. 12
Check to see that students color the correct bead to show the ordinal number.

p. 13
1. count on: 6; sum: 6
2. count on: 4, 5; sum: 5
3. count on: 8, 9; sum: 9
4. count on: 5, 6, 7; sum: 7

p. 14
Check to see that students draw the correct number of dots.
1. 3 + 3 = 6
2. 5 + 5 = 10
3. 2 + 2 = 4
4. 4 + 4 = 8
5. 6 + 6 = 12
6. 1 + 1 = 2

p. 15
Check to see that students draw the correct number of Xs.
1. 4; 4
2. 6; 6
3. 5; 5
4. 5; 5

p. 16
Check to see that students draw the correct number of circles.
1. 2
2. 3
3. 5
4. 1
5. 6
6. 4

p. 17
Check to see that students draw the correct number of Xs.
1. 0
2. 3
3. 5
4. 0
5. 0
6. 7

p. 18
1. 5 + 2 = 7; 2 + 5 = 7; 7 − 2 = 5; 7 − 5 = 2
2. 7 + 1 = 8; 1 + 7 = 8; 8 − 1 = 7; 8 − 7 = 1

p. 19
1. even
2. odd
3. even
4. odd
5. odd
6. even

p. 20
1. 5 + 4 = 9; 6 + 3 = 9; 7 + 2 = 9; 8 + 1 = 9
2. 4 + 4 = 8; 5 + 3 = 8; 6 + 2 = 8; 7 + 1 = 8
3. 3 + 4 = 7; 4 + 3 = 7; 5 + 2 = 7; 6 + 1 = 7
4. 2 + 4 = 6; 3 + 3 = 6; 4 + 2 = 6; 5 + 1 = 6

p. 21
Check students' tables.
Answers to question will vary.
Accept reasonable responses.

p. 22
1. 10, 20, 30, 40, 50, 60, 70
2. 10, 20, 30, 40, 50, 60, 70, 80
3. 10, 20, 30, 40, 50, 60, 70, 80, 90

p. 23
6; 10; 14; 20; 22; 24; 28; 36; 40; 42; 44; 48; 50

p. 24
20; 30; 35; 45; 60; 65; 70; 80; 90; 95

p. 25
1. Students show 1:00.
2. Students show 5:00.
3. Students show 11:00.
4. 6:00
5. 4:00
6. 2:00

p. 26
1. Students color the 7th, 14th, 21st, and 28th red.
2. Students color the 3rd blue.
3. Students color the 27th yellow.
4. Students color Tuesday the 18th green.
5. Check that the 12th and 22nd are orange.

p. 27
Check students' patterns.

p. 28
1. d, e, f
2. q, r, s
3. vase
4. rope
5. cat
6. cats or kittens
7. jumps
8. runs

p. 29
1. f, g, h
2. n, o, p
3. c, d, e
4. j, k, l
5. q, r, s
6. w, x, y

p. 30
1. bike
2. tape
3. tube
4. robe
5. rake
6. kite

p. 31
1. two dogs
2. one rabbit
3. one cow
4. four cats

p. 32
1. swims
2. swings
3. plays
4. talks

p. 33
fly; rat; mouse

p. 34
1. Students circle 9 pennies.
2. a farm
3. Order may vary: East, West.
4. Students draw a farm symbol any place west of the castle.

p. 35
1. Answers will vary.
2. Possible answer: The principal makes sure the rules of the school are followed.
3. Answers will vary.
4. Possible answer: The teacher keeps the class together so students can learn.

p. 36
Check students' maps and map keys. Most students will use the same symbols in the key as on the map.

p. 37
1. East; shaded triangle
2. East; white triangle
3. West; shaded triangle
4. South; white triangle

p. 38
First: D
Second: A
Third: C
Fourth: B

p. 39
1. Students circle 7 pennies.
2. Students circle 9 pennies.
3. Students circle 1 nickel.
4. Students circle 5 nickels.

p. 40
Picture number order: 3, 1, 4, 2
1. seed
2. bird
3. gloves
4. clouds

p. 41
Order: 4, 1, 5, 3, 2, 6

p. 42
1. seed
2. root
3. flower
Students paste the pictures in this order: seed, roots, stem, leaves.

p. 43
1. S
2. F
3. W
4. W
5. W
6. F
7. F
8. W
9. S

p. 44
1. winter
2. spring
3. summer
4. fall

p. 45
All 4 pictures in the top row belong with the sunny picture on the left. All 4 pictures in the bottom row belong with the rainy picture on the right.